SYNONYMS AND ACRONYMS IN AGRICULTURE AND FORESTRY

About the Author

Professor Dr. O. Anchanam Alagia Pillai was educated at the Tamil Nadu Agricultural University, Coimbatore, Tamil Nadu, India, and College of Forestry, University of Idaho, Moscow, USA. He is currently working as Head, Horticultural Research Station, Tamil Nadu Agricultural University, Pechiparai-629 161, Tamil Nadu, India. He has had extensive, research education and extension experience and has devoted his 34 years of professional life to Tropical and Temperate Horticulture. He has published 14 books in Tamil, including 3 English-Tamil Dictionaries on Agriculture, 50 research articles, and contributed chapters in many books on Horticulture and Forestry. He is also a versatile speaker and has participated in 50 farmer's functions so far.

He has been engaged in undergraduate and Postgraduate teaching and research in Horticulture and Forestry and guided many postgraduate students. Professor Dr. O. Anchanam Alagia Pillai has received many honours and award also. These include an award from Govt. of Tamil Nadu for writing a book on Forestry.

SYNONYMS AND ACRONYMS IN AGRICULTURE AND FORESTRY

Dr. O. Anchanam Alagia Pillai
Professor and Head
Horticultural Research Station
Tamil Nadu Agricultural University,
Pechiparai (Tamil Nadu)

Foreword by
S. Sankaran
Vice-Chancellor
Tamil Nadu Agricultural University,
Coimbatore (India)

CONCEPT PUBLISHING COMPANY, NEW DELHI-110059

ISBN 81-7022-693-7

First Published 1999

Published and Printed by

Ashok Kumar Mittal
Concept Publishing Company
A/15-16, Commercial Block, Mohan Garden
New Delhi-110059 (India)

Phones: 5648039, 5649024
Fax: 091-(11) 5648053

FOREWORD

The Agriculture Science is growing with many new findings. Our country is fortunate in its agro-ecological endowments. We have over 20,000 plant species occurring in our Country, a number which is far higher than those found in countries with a larger land mass. We have over 4000 woody species of plants and among them about 50 are of major utility. The potential productivity of our forests is also higher.

The study of Agriculture in the World is much respected from 1960 onwards. We need many text books and reference books to study and do research in various problems of Agriculture, Horticulture and Forestry.

Dr. O. Anchanam Alagia Pillai, Professor of Horticulture and Forestry, who was also a Scientist trained in U.S.A. He has come forward to bring out a reference book with details for the unknown Synonyms and Acronmys prevalent in the Literature of Agriculture, Horticulture and Forestry through Concept Publishing Company, New Delhi.

This book contains details for more than 1000 Synonyms and Acronyms used in Agriculture and Forestry. The author took pain to collect valuable details and bring out a nice book.

This book will be very useful to the students and scientists of Agriculture, Horticulture and Forestry in the Agricultural Universities and

Institutions. Definitely this will serve as a good reference book for the extension people of Agriculture also. The general people may also get some information for knowing the meaning of Synonyms and Acronyms.

I congratulate the author Dr. O. Anchanam Alagia Pillai, Professor and Head of Horticulture, Tamil Nadu Agricultural University, Coimbatore for his effort to bring out this valuable book.

July 3, 1996

S. SANKARAN
Vice Chancellor
Tamil Nadu Agricultural University
Coimbatore-641003 (India)

PREFACE

Indian Agriculture after many metamorphoses has came forward successfully to provide food to the entire population of India after passing through many hurdles and struggles. Eventhough nature and some policies in agriculture are not supportive, the Agricultural Scientists and Extension Workers are working hard every day to meet the food demand of the people with many fascinating achievements in the food front compared to other world countries.

These achievements are possible through the development of high yielding and hybrid varieties in most of the Agricultural and Horticultural crops, advancement of regional management of agricultural practices after NARP programmes along with T&V and other World Bank aided programmes, cost effective and environment cleaning IPM programmes, use of micronutrients, Bio-fertilizers etc.

The export potentialities of horticultural crop has paid rich dividend in Indian Economy. But we need separate collection and Export Departments with huge transporting facilities for export of fruits, vegetables, spices, flowers and medicinal products, with world wide Indian selling agencies. The Agricultural and Horticultural Education also needs to be reoriented with emphasis on Export and self-employment by establishing Agricultural and Horticultural Education Control Boards.

The Forest programmes are also still under protective and conservation styles. The natural calamities, policies like Land Patta Distribution Programmes and Intensive Agricultural Programmes have taken away large chunk of national forest areas and depleted the forest acreage.

It is a good sign that some of the Agricultural Universities have started Forest Education Programmes during 1980's and provided forest graduates for Forest Department to strengthen their working force with latest Academic standards.

In addition to successful Forestry and Agro-Forestry Programmes the waste land improvement programmes also emerged as a profitable programmes with planting of Teak, Acacia, Mangium, Eucalyptus, Albizia and other species in non-forest waste land. These private green forest programmes needs to be watched and governed with strict government rules and laws also.

Hitherto the literature published about Indian Agriculture followed conventional standards with review patterns. During the last 10 years India witnessed a good change among Agricultural Scientists and Professors to write and publish Dictionaries on various branches of agriculture to throw meaningful light on every concept and word used in Agriculture.

In continuing the foot steps this new volume "Synonyms and Acronyms in Agriculture and Forestry" is written and published through leading Concept Publishing Company, New Delhi.

This volume provides latest information about innumerable Government and private organizations engaged in Agriculture and Forestry, various terminology and concepts used in different branches of Agriculture and Forestry Sciences alphabetically. The aim is to educate and simplify the short words. The target people are students, scientists, and extension workers.

I record my humble and sincere thanks to our respectful Vice-Chancellor, Dr. S. Sankaran for very kindly consenting to provide a Foreward for this book.

I would like to record my appreciation of the valuable suggestion and help, I had from Thiru J.C. Pillai of Tamil Nadu Book House, Madras.

Finally I must gratefully thank the renowned Publisher, Concept Publishing Company, New Delhi for undertaking the Publication and for bringing out the Dictionary in time and in excellent form.

DR. O. ANCHANAM ALAGIA PILLAI

SYNONYMS AND ACRONYMS IN AGRICULTURE AND FORESTRY

A

a : are (10^2 m^2); foliage area density (m^2 m^{-2}); number of acres; radius of Bohr orbit (Radio isotope); acceleration (mechanical unit)

Å : Angstrom (atomic scale)

A : Argon area (m^2); ampere; Aggressivity

A : Disintegration rate (= -dn/dt); Mass number (Radio Isotope)

A (horizon) : The uppermost horizon of the mineral soil; essentially a horizon of eluviation (washing out)

Å (horizon) : Humus layers consisting of mull and mor, the latter frequently composed of F and H layers

$\ddot{\mathring{A}}$ (horizon) : Surface litter

A_1 horizon : Dark layer with high content of organic matter

A_2 horizon : Light coloured zone of maximum eluviation

A_3 horizon : Transitional zone to B, but more akin to A

A_A : Allowable cut derived Variable land production function

AAN : The American Association of Nursery man, (Washington D.C., USA)

AARRO : Afro-Asian Rural Reconstruction Organisation

AAS : Atomic Absorption Spectrophotometer

AASCO : Association of American Seed Control Officials (USA)

AAU : Assam Agricultural University, Jorhat-785013, Assam, radia

AAZP : Arignar Anna Zoological Park, Chennai, India

ABA : Abscisic Acid

ABI : American Biographical Institute, North Carolina, USA.

abs : Absolute

ABS : American Bamboo Society

A.C. : alternating current

ACDEP : Association of Church Development Programme/Tamla, Ghana.

ACTS : African Centre for Technology Studies, P.O. Box. 45917, Nairobi, Kenya.

A/c Horizon : The mid-horizon of recently developed generally calcareous soils, in which the quantity of unweathered rock increases with depth to merge into unweathered parent material.

Ac : Actinium (element)

ACCF : Additional Chief Conservator of Forests

ACE : Allowable Cut Effect

ACF : Assistant Conservator of Forests

ACIAR : Australian Centre for International Agricultural Research, Canberra, Australia

ADB : Asian Development Bank (Manila, Philippines)

ADF : Acid detergent fibre

ADP : Adenosine diphosphate

ADR : Associate Director of Research

a.e. : Acid equivalent

AE : Efficiency derived variable land production function

AED : Academy for Educational Development, Washington DC, USA.

AERB : Atomic Energy Regulatory Board

AF : Idle (fallow acres)

AFF : Africa Enterprise Fund (Financed through IFC)

A/F ratio : Abundance to frequency ratio

AFNETA : Alley Farming Network for Tropical Africa (Ibadan, Nigeria)

AFNOR : Association Francaise de Normalisation

AFOCEL : Association Foret Cellucose, (Paris, France)

AFRC	: Agricultural Food Research Council, U.K.
AFRENA	: Agro-Forestry Research Network for Africa (of ICARF)
AFTA	: Association for Temperate Agro-Forestry, Michigan State University, USA
Ag	: Acres of old growth; Silver (element)
AGR	: Absolute growth rate
A Grade thinning	: Light thinning
AGRIS	: Agricultural Research Information System (FAO, Rome, Italy)
AIDAB	: Austrialian International Development Assistance Bureau
a.i.	: active ingredient
A.I.	: Artificial Insemination; Aridity index; Activity index; Artificial intelligence
A.I.D	: Agency for International Development
A.I.F.C.	: Associate of Indian Forest College
AIFM	: ASEAN Institute for Forest Management, Kualalumpur, Malaysia
AIT	: Asian Institute of Technology, GPO Box. 2754, Bangkok-105 of Thailand
AK	: Altocumulus (cloud)

Al : Aluminium (element)

alt. : Altitude

Am : regulated (managed) acres

AMAI : Alkali Manufacturer's Association (New Delhi)

AMP : Adenosine Monophoshate

Amu : Atomic mass unit

Anat. : Anatomy

AND : Anderson Medium

ANOVA : Analysis of variance

ANU : Australian National University (Canberra, Australia)

AO : Andaman Ordianary (Coconut variety)

AOAC : Association of Official Analytical Chemists, USA

AOSCA : Association of Official Seed Certifying Agencies

APAN : Asia Pacific Agroforestry Network

APAU : Andhra Pradesh Agricultural University, Hyderabad-500 030. INDIA

APEC : Asia Pacific Economic Cooperation

APO : Annual Plan of Operation

APPLE : Aerian Passenger Pay Load Experiments

Approx. : Approximately

APSW : Awadhesh Pratap Singh University, Rewa, M.P., India

Ar : residual acres; Argone (element)

ARC : Arasu Rubber Corporation, Nagercoil, Tamil Nadu, India

Archaeol : Archaeology

ARIC : Agricultural Research Information Centre

Arith : Arithmatic

ARMEF : Association of Pourla Rationalisation et le mecanisation de exploitation Forestiere

ARS : Agricultural Research Station; Agricultural Research Service

As : Soil specific gravity

As : Arsenic (element)

AS : Altostratus (cloud); Ammonium Sulphate

Asean : Association of South-East Asian Nations

ASD : Average Stand Diameter

ASLV : Augmented Satellite Launch Vehicle

ASM : American Society of Metals

ASOCON : Asia Soil Conservation for the Humid Tropics Network

ASRB : Agricultural Scientists Recruitment Board, Pusa, New Delhi

ASTA : American Seed Trade Association Washington D.C., USA

ASTM : American Society for Testing Materials

Astron : Astronomy

At : Astatine (element)

ATC : Agreement on Textiles and Clothing

ATM : A Synchronous Transfer Model

atm : (standard) atmosphere

at.no. : atomic number

at.wt. : atomic weight

ATP : Adenosine triphosphate

Au : Gold (element)

Au : (i) Annamalai University, Chidambaram, Tamil Nadu, INDIA.
(ii)Andhra University, Visakhapatnam, INDIA.

AUM : Animal Unit Month

av : average

AVRDC : Asian Vegetable Research and Development Centre, Shanhua, Taiwan, Republic of China.

AZO : Azospirillum

B

b : Wien's displacement law constant (= 0.2898 cm deg); Barn

b.p. : Boiling point

B : Boron (element); Binding energy; Buildup factor (Radio isotope);

Magnetic flux density (Electrical unit)

B horizon : Essentially an illuvial horizon enriched with material washed down from A horizon and generally much more compact.

B1 horizon : Transitional zone to B, but more akin to B

B2 horizon : Maximum accumulation of silicate clay minerals or of iron and organic matter

B3 horizon : Transitional zone to C but more akin to B

b.a. : basal area

Ba : Barium (element)

BAI : area index of woody canopy elements (poles, branches, twigs etc.) (m^2 m^{-2})

BAIF : Bharatiya Agro Industries Foundation, Pune, (India)

BALCO : Bharat Aluminium Company, India

BARC : Bhabha Atomic Research Centre, Trombay, Mumbai, Maharashtra, India

BAU : Birsa Agricultural University, Kanke, Ranchi 834 006, Bihar, India.

BBMV : Banana Bract Mosaic Virus

B.B.S. : Bharatpur Bird Sanctuary, India

BC_1, BC_2 : The first back cross generation; second back cross generation, etc.

BCA : Benefit Cost Analysis

BCKV : Bidhan Chandra Krishi Vishwa Vidyalaya, Kalyani, W.Bengal, India

BCR : Benefit Cost Ratio

bd. ft. : board feet

BDS : Bio Digested Slurry

Be : Beryllium (element)

BER : Breadthwise Elongation Ratio

B.F.F.System : Broad bed and furrow System

BGA : Blue Green Algae

B Grade thinning : moderate thinning

b.h. : breast height

Bi : Bismuth (element)/regression coefficient

Bio : Bose Institute of Oceanography, Calcutta, India

BIOSIS : Bio Science Information Service

BIS : Bureau of Indian Standards

BM : Bordeaux Mixture

BNHS : Bombay Natural History Society, Mumbai, India

BNL : Brookhaven National Laboratory, USA

BOD : Biological Oxygen Demand

BOLT : Bombay on line Trading

BOSTID : Board on Science and Technology for International Development, O/o. International Affairs, National

Research Council, Washington DC, USA.

bp : base pairs

B.P. : Boiling Point; Before Present; British Pharmacopia

BPH : Brown Plant Hopper

Br : Bromine (element)

BS(I) : British Standard (S) Institution

BSI : Botanical Survey of India, Calcutta, India

BSOM : Benzene Soluble Organic Matter

BSV : Banana Streak Virus

BTCV : British Trust for Conservation Volunteers

Btu : British Thermal Unit

BTV : Bunchy Top Virus

bu : Bushel

BWR : Boiler Water Reactor

BY : Biological Yield

C

c : centi (X 10^2); Velocity of light (Radio isotope); About or approximately

C : Cost; Carbon (element); Centigrade; Coulomb; Concentration, Capacitance (Radio Isotope); Cirrus (cloud)

C (horizon) : The uppermost layer of parent material which is in process of forming true soil under the influence of weather

°C : degree celsius

C1 : First radiation constant (=4.992 × 10^{-5} erg cm)

C2 : Second radiation constant (=1.439 cm deg)

Ca : Calcium (element)

CA : Controlled Atmosphere

CAB : Commonwealth Agricultural Bureaux (International since 1986) (Farnham Royal, England)

CABO : Centre for Agrobiological Research, Wagesingen, Netherland.

CAC : Committee on Capital Account Convertibility

CACP : Commission for Agricultural Costs and Prices

CAD : Computer Aided Design

CADP : Command Area Development Programme

CAE : Coordinated Agronomical Experiment

CAF : Chinese Academy of Forestry (Beijing, China)

c.a.i. : current annual increment

Cal. : Calorie

CAM : Crassulacean Acid Metabolism

CAMP : Cyclic Adenosin Monophosphate

CARE : Cooperative for American Relief for Everywhere

CARI : Central Avian Research Institute, Izatnagar, K.P., India

CARIANGI : Central Agricultural Research Institute for Andaman and Nicobar Groups of Islands, Port Blair, Andaman and Nicobar 744 101.

CAS : Controlled atmospheric storage

CASLE : Commonwealth Association of Surveying and Land Economy

CATIE : Centro Agronomico Tropical de Investigaciony; Costa Rica Ensenanza, Twrialla

CAU : Central Agricultural University, Imphal, India.

CAZRI : Central Arid Zone Research Institute, Jodhpur-342 003 (India)

Cb : Columbium

CB : Coir Board, Kochi, Kerala, India; Coffee Board, India.

CBD : Convention of Biological Diversity (Signed in Rio in 1992)

CBIP : Central Board of Irrigation and Power, India.

CBN : Cubic Boron Nitride

CC : Cell constituents/Cochin China (Coconut variety)

CCA : Copper Chrome Arsenate

CCDP : Combined Cycle Demonstration Plant

CCDS	: Centre for Development Cooperation Services, Amsterdam, Netherlands.
CCF	: Chief Conservator of Forest; Crown Competition Factor
CCI	: Cotton Corporation of India, Mumbai, India.
CCMB	: Center for Cellular and Molecular Biology, Hyderabad, A.P., India
CCS	: Commercial Care Sugar Percent; Common Channel Signalling
CCSHAU	: Chaudhary Charan Singh Haryana Agricultural University, Hissar 125 004, India
CCT	: Correlated Curve Trend
CCTA	: Comision de Coordinacin de Tecnologa Andina, Lima, Peru.
CD	: Critical difference/C zapek-dox (growth medium)
Cd.	: Candela; drag Coefficient (dimensionless); Cadmium (element)
CDB	: Coconut Development Board, Kochi, Kerala, India
CDO	: Chowghat Dwarf Orange (Coconut Variety)
CDSU	: Classification and Diagnostics of soils of the USSR
C/D thinning	: Intermediate grades of thinning between C and D.

CDR : Centre for Development Research, Copenhagen, Denmark.

CDRI : Central Drug Research Institute (Lucknow, India)

CDTF : Community Development Trust Fund, Dares Salaam, Tanzania.

Ce : Cerium (element)

CEC : Cation exchange capacity

CEE : Centre for Environmental Education (Ahamedabad, Bangalore)

CERI : Central Food Technological Research Institute, Mysore, India

CERPA : Centre for Research Planning Action, New Delhi.

CERT : Centre for Ecological Research and Training, (Bangalore), India

CF : Chief Conservator of Forests; Crude Fibre

CFA : Commonwealth Forestry Association (Oxford, England)

CFB : Corrugated Fibre Board

CFC : Central Forest Commission, Chloro-fluro carbon

CFI : Commonwealth Forestry Institute (Oxford, England)

CFRC : Central Forest Rangers College, Chandrapur (Maharashtra) India

CFRI : Central Fodder Research Institute, Jhansi, India.

CFSC : Community and Family Study Center (University of Chicago)

CFTC : Commonwealth Fund for Technical Cooperation

CFTRI : Central Food Technological Research Institute, Mysore, India.

Cg : Centigramme

CGIAR : Consultative Group of International Agricultural Research, C/o World Bank, Washington, USA, F.A.O., Rome.

CGR : Crop growth rate; Crop Genetic Resource

Cgs : Centimeter-gram-second

C grade thinning : Heavy thinning

CGWB : Central Ground Water Board

Ci : Curie

CI : Chlorine (element)

CI : Conservation International

CIAE : Central Institute of Agricultural Engineering, Bhopal 462 003. H.P., India

CIAT : Centro Internacional de Agricultura Tropical, Cali (Colombia)

CIBA : Central Institute of Brackish Water Aquaculture, Chennai-500 028, Tamil Nadu., India

CIBC : Common wealth Institute for Biological Control, Trinidad, West Indies.

CICFRI : Central Inland Capture Fisheries Research Institute, Barrackpore 743 101, W.B., India

CICR : Central Institute for Cotton Research, Nagpur 440 001, Maharashtra, India

CIDA : Canadian International Development Agency, Cubac.

CIDICCO : International Cover crop Clearing House, Tegucigalpa Dc, Honduras

CIFA : Central Institute of Fresh Water Aquaculture, Bhubaneshwar 751 002 Orissa, India

CIFE : Central Institute of Fisheries Education, Mumbai.

CIFT : Central Institute of Fisheries Technology, Willingdon Island, Cochin, Kerala, India

CIHNP : Central Institute of Horticulture for Northern Plains, Lucknow, U.P'., India

CII : Cropping Intensity Index/Confederation of Indian Industry

CIKARD : Centre for Indigenous Knowledge for Agriculture and Rural Development, 10WA, USA.

CIL : Coal India Ltd., Calcutta, India.

CILSS : Comite Permanent inter—elats de Lutte Countre fa Secheresse dans de Sahel

CIMMYT : Centro Internacional de Mejoramiento de Maizy Trigo, Mexico.

CIP : Central Issue Price; Centro Internacional de la Papa, Lima, Peru, South America.

CIRDAP : Centre on Integrated Rural Development for Asia and Pacific, New Delhi.

CIRG : Centre Institute for Research on Goats, Makhdoon, Mathura, India.

CITES : Convention on International Trade in Endangered Species of wild Fauna and Flora.

CK : Cirrocumulus (cloud)

CL : Common Lands

CLI : Central Lac Research Institute, Ranji, India.

CLRI : Central Lather Research Institute, Madras, Chennai, India.

CLUI : Cultivated land utilization index

Cm^2 : Square centimetre

Cm^3 : Cubic centimetre

Cm : Centimetre

CME : Centre for Mining Environment Dhanbad, India

CMERI : Central Marine Fisheries Research Institute, Kochi, Kerala, India

CMFRI : Central Marine Fisheries Reserch Institute, Kochi, Kerala, India,

CMRS : Central Mining Research Station

C-Met : Centre for Materials for Electronics Technology, Pune, Hyderabad, Thrissur, India.

CMS : Cytoplasmic male sterility

CNN : Cable News Network

CNP : Corbett National Park, U.P., India.

C:N ratio : Carbon-Nitrogen ratio

CNRS : Centre National de la Researche Scientifique (France)

CNS : Central Nervous System

CNSL : Cashewnut Shell Liquid

Co : Cobalt (element), Carbon monoxide

CON : Centre for Development Work, Wageningen, Netherlands.

Collpro : Collaborative Program (or ICRAF)

COOIT : Central Organization for oil Industry and Trade

COD : Chemical Oxygen Demand

COEC : Commission of European Communities

Coeff. : Coefficient

COMECON : Council for Mutual Economic Aid

Conc. : Concentrated

COP : Conference of Parties (Earth Summit)

Coul : Coulomb

CP : Specific heat of air at constant pressure (J kg^{-10} k-1); Commercial product/Commercial Paper

CP : Crude Protein; chemically pure

CPA : Critical Path Analysis

CPCB : Central Pollution Control Board, New Delhi

CPCRI : Central Plantation Crops Research Institute, Kasargod, Kerala., India.

CPE : Cumulative Pan Evaporation

CPI : Consumer Price Index

cpm : counts per minute

CPM : Critical Path Method

C.P.P. : Crash Plantation Project

C.P.O.A.C. : Critical Operating Production Allocation Control

CPPRI : Central Pulp and Paper Research Institute, Saharanpur, U.P., India.

CPRI : Central Potato Research Institute, Shimla-171 001, H.P., India.

C.P.S. : Critical Path Schedule; counts per second

Cr : Chromium (element)

CR : Conception Rate/Competition Ratio

CRBD : Completely Randomised Block Design

CRC : Coniferous Research Centre, Shimla, India.

CRD : Completely Randomized Design

CRIDA : Central Research Institute for Dryland Agriculture, Hyderabad, A.P., India.

CRM : Counting Rate Meter

CRR : Cash Reserve Ratio

CRRI : Central Rice Research Institute, Cuttack, Orissa, India.

CRS : Cumulative Recovery System

CRSARD : Centre for Research on Sustainable Agriculture and Rural Development

CRZ : Coastal Regulation Zone

Cs : Caesium (element)

CS : Cirrostatus (cloud)

CSD : Commission on Sustainable Development

CSD : Centre for Science and Development, New Delhi-110 019, India.

CSE : Centre for Science and Environment

CSIR : Council of Scientific and Industrial Research, New Delhi.

CSIRO : Commonwealth Scientific and Industrial Research Organization (Canberra, ACT 2600, Australia)

CSSRI : Central Soil Salinity Research Institute, Karnal, Haryana-132 001, India

CSWCRTI : Central Soil and Water Conservation Research and Training Institute, Dehradun, U.P., India.

CSWRI : Central Sheep and Wool Research Institute, Avikanagar 304 501, Rajasthan, India.

CT : Commercial thinning volume/ Carbon Tetra chloride

CTCRI : Central Tuber Crops Research Institute, Trivandrum 695 017 Kerala, India.

CTD : Centre for Technology Development, Bangalore.

ctl : Central

Ct DNA : Chloroplast DNA

CTET : Central Technique Forestier Tropical (France)

CTRI : Central Tobacco Research Institute, Rajamundry, A.P., India

Cu : Cubic; Copper (element)

Cunit : 100 cubic feet

CUSO : Canadian University Services Overseas, Ottawa, Ontario, Canada.

cv : Cultivar

CV : Coefficient of Variation

CWLW : Chief Wild Life Warden

CWRDM : Central Water Resources and Drainage Management

C.W.S. : Coppice-with-standard

cwt : hundred weight

D

d : day; zeropene displacement (m); distance, cm; Deuteron (Radio isotope)

D : Diffusion Coefficient; demand

D(trees) : dominated trees (class symbol II)

D1 : Predominant trees

D2 : Codominant trees

DANIDA : Danish International Development Agency

DAP : Dihydroxy Acetone Phosphate; Di Ammonium Phosphate

DAR : Days after ratooning

DAS : Days after sowing

dB : decibel

d.b.h. : diameter at breast height

DBT : Department of Bio Technology, New Delhi, India.

D.C : direct current

DCF : Deputy Conservator of Forests

DCFRN : Developing Countries Farm Radio Network, Toronto, Ontaria, Canada.

DCP : Digestible Crude Protein

DCR : Desilting cum Reclamation of Tanks.

D & D : Diagnosis and Design

DD : Deputy Director

DDM : Digestible Dry Matter

DDT : Dichloro-Diphenyl Trichloro-Ethane

DE : Digestible Energy; Dose equivalent

D/E : Intermediate grades of thinning between D and E

DFO : District Forest Officer

DGI : Dickson Quality Index

D Grade thinning : Very heavy thinning

DI : Diversity index

DIALOG : Lockhood DIALOG Computer retrieval System (California, USA)

Diam : Diameter

DIGF : Deputy Inspector General of Forest

dil. : dilute

DIMDI : Deutsches Institute for Medizinische Dokumentation and Information, Koln, Germany

DIN : Deutshes Institute for Normung

DIPAS : Defence Institute of Physiology and Allied Sciences

DI.F.O. : Divisional Forest Officer

DLM : Directorate of Lac Development Ranchi, India.
DM : Dry Matter; Divisional Manager
DMRT : Duncan's Multiple Range Test
dl : decilitre
DNA : Deoxyribosnuecleic Acid
dob : Diameter over bark
DOM : Digestible Organic Matter
DPD : Diffusion Pressure deficit
DPGA : Diphospho Glyceric Acid
dpm : Disintegrations per minute
DPPQ&S : Directorate of Plant Protection, Quarantine & Storage, Faridabad
dps : Disintegrations per second
DRDA : District Rural Development Agency
DST : Department of Science and Technology, India
d.u.b. : Diameter under bark
DW : Dry weight
Dy : Dysprosium (element)
dyn : Dyne
D.Z.P. : Director of Zoological Park

E

e : Vapour pressure (K Pa); Regeneration category established; Base of Naperian (natural) logarithms; Elementary charge; Electric charge (Radio Isotope)

E	: einstein; flux density of water vapour in air positive upwards g m^{-2} s^{-1}; East; Silvicultural effort; Total energy (Radio Isotope) Single Electrode Potential; Potential (Electrical unit)
EAO	: Ecological Agricultural Projects, Quebec, Canada.
EAPI	: East-west Environment and Policy Institute, (Hawaii)
e.c.	: emulsifiable concentrate;
EC	: European Communities; Electrical Conductivity; Electron Capture
ECAFE	: Economic Commission for Asia and Far East
ECE	: Economic Commission for Europe (UN)
ECP	: Exchangeable Cation percentage
EDI	: Electronic Data Interchange
EDTA	: Ethylene Di Amin Tetra Acidic Acid
EEC	: European Economic Community
EEM	: Efficient Environmental Management
EFRC	: Eastern Forest Ranger's College, Kursheong, Assam, India
e.g.	: For example
EGC	: Epigallo Cathechin (Tea Polyphenol)

EGCA : Epigallo-Cathechin-3 gallale (Tea Polyphenol)

E Grade thinning : Very heavy thinning

ELCI : Environment Liasion Centre International, Nairobi, Kenya.

ELISA : Enzyme-linked Immunosorbent Assay

EM : Electron Microscope

EMBRAPA : Brazilian Agricultural Research Company

EMS : Ethyle methane sulphonate

e.m.c. : equilibrium moisture content

emu : Electro magnetic unit

ENSA : Ecole Nationale Superieur, Agronomique

Eo : Open pan evaporation

Ep : Elasticity of Production

EPA : Environment Protection Agency, USA.

EPIP : Export Promotion Industrial Park

equiv : Equivalent

Er : Erbium (element)

ER : Endoplasmic reticulum

ERA-IRS : European Space Agency—Informational Retrieval System (Italy)

ES : Export System

ESCAP : United National Economic and Social Commission for Asia and Pacific

ESP : Exchangeable Sodium Percentage; Extra Sensory Perception

ESR : Electron Spin Resonance

es (T) : Saturation Vapour Pressure at Temperature T (K Pa)

esu : Electrostatic unit

et al. : et alii (and others)

ET CROP : Crop evapotranspiration

ET50 : Median effective time

Eu : Europium (element); European Union

eV : Electron Volt

F

f : Frequency

F : Forma (form); harvest volume; Fluorine (element); Filial generation; Farad; Force (Mechanical unit); Inbreeding coefficient

°F : degree Fahrenheit; Force

F1 : First Filial Generation after a cross

F2 : Second Filial Generation obtained by self fertilization or crossing inter se of F1 individuals

F3 : Progeny obtained by self fertilizing F2 individuals.

FAD : Flavour Adenine Dinucleotide

FAO : Food and Agricultural Organisation Rome, Italy.

FAQ : Fair Average Quality

Fe : Flux of CO_2 in air positive upwards (gm^{-2} s^{-1})

FC : Field capacity; Felling cycle

FCM : Fat-Corrected Milk

FDP : Fructose-Diphosphate

Fe : Iron (element)

FE : Forest Economist

FEMAS : Formulation Evaluation Monitoring and Statistics

FERRO : Far Eastern Regional Research Office of the United States Department of Agriculture

FFA : Free fatty acids

FFD : Fractional Factorial Design

FICCI : Federation of Indian Chambers of Commerce and Industry

FGNFTS : Fast Growing Nitrogen Fixing Trees

floz : fluid ounce

FMC : Forward Market Commission

FOE : Friends of Earth

FOGIA : French Oil and Gas Industry Association

FPC : Finite Population Correction (Statistics)

FPB : Floating Periodic Block

Fr : Francium (element)

FRAME : Fund for Replacement of Animals in Medical Experiments

FRI : Forest Research Institute (Dehradun, India)

F.S. : Felling Series

FSI : Forest Survey of India, Dehradun

FSM : Forest Service Manual

FSR : Farming System Research

F.P.S. : Forest Protection Squad

f.sp. : forma specialis (special forms)

ft. : foot (feet)

FTGR : Forest Tree Genetic Resource

F.U. : Feed Unit; Fodder Unit

F.U.O. : Forest Utilization Officer

FV : Feeding value

FW : Fresh Weight

FYM : Farm Yard Manure

G

g. : acceleration due to gravity; gramme

gn : gravity; standard

G : flux density of heat in soil positive downwards (W m^{-2}); Physical geometry factor (Radio Isotope)

G (horizon) or Gley horizon : This occurs between B and C horizon in soils with imperfect or impeded drainage

Ga : Gallium (element)

GA : gibberellic Acid

Gal. : gallon (imperial)

GATT : General Agreement on Tariffs and Trade

GAU : Gujarat Agricultural University Sardar Krushi Nagar, Banaskantha-385 506, India.

GBPUAT : Govind Ballabh Pant University of Agriculture and Technology, Pant Nagar, U.P.-263 145, India.

g.p.h. : girth at breast height

Gc : Gas Chromatography

GCA : General Combining Ability

Gd : Gadolinium (Element)

GDP : Gross Domestic Product

Ge : Germanium (element)

G.E.R.T. : Graphical Evaluation and Review Technique

GEMS : Global environmental Monitoring System (UN sponsored)

gen.nov. : genus novum (new genus)

GLC : Gas-Liquid Chromatography

G.M. : General Manager

GNP : Gross National Product

G.O.I. : Government of India

GOL : Grain-Oil seed-livestock model

GRAIN : Genetic Resources Action International, Barcelona, Spain.

Gram+ : Gram Positive

Gram– : Gram Negative

GR : Genetic Resources

G.S. : Growing Stock

GSDP : Gross State Domestic Product

GSI : Geological Survey of India

GSV : Grassy Stunt Virus

Gy : Gray

H

h. : hour; height of plants (m); Planck's constant (Radio Isotope)

h2 : heritability

H : Hydrogen (element); flux density; henry; Magnetic field strength (electrical unit)

ha : hectare

HAU (eeshau) : Haryana Agricultural University, Hissar

H.B.N.P. : Hazari Bagh National Park, Bihar, India

H.C.grade thinning : Heavy crown thinning

HCH : Benzene Hexachloride

HCT : Hematocrit

HDI : Harvest Diversity index

He : Helium (element)

Hf : Hafnium (element)

Hg : Mercury (element)

H.I. : Harvest Index

hkg : hectokilogram

hl : hectolitre

HLRF : Hindustan Lever Research Foundation

Ho : Holmium (element)

hp : Horse power

HPKV : Himachal Pradesh Krishi Vishwa Vidyalaya, Palampur-176 062, India.

HPLC : High Performance Liquid Chromatography

hr : Hour

ht : height

HVL : Half value layer

Hz : hertz

I

I : gross interception of rainfall (m); Iodine (element); Intensity of Radiation (Radio Isotope); Current (electrical unit)

IAA : Indole Acetic Acid

IAEA : International Atomic Energy Agency, Vienna.

IARC : International Agricultural Research Centre

IARI : Indian Agricultural Research Institute, New Delhi, India.

IASA	: International Alliance for Sustainable Agriculture, University of Minnesota, Minneapolis, Minnesota, USA.
IASCP	: International Association for the Study of Common Property, U.S.A.
IASRI	: Indian Agricultural Statistics Research Institute, New Delhi-110 012.
IAZFR	: Institute of Arid Zone Forestry Research, Jodhpur, India.
IB	: Internal Bond
IBA	: Indolebutyric Acid
IBGE	: Instituto Brasiliero de Geografia e Estadistica (Brazil)
IBPGR	: International Bureau for Plant Genetic Resources (C/o FAO, Rome, Italy)
IBRD	: International Bank for Reconstruction and Development (Washington, D.C.U.S.A. (now World Bank)
IBSNAT	: International Benchmark Soils Network for Agrotechnology Transfer, Hawaii.
IBSRAM	: International Board for Soil Research and Management, Bangkok, Thailand.
ICAN	: International Conservation Action Network

ICAR : Indian Council for Agricultural Research, New Delhi-110 001.

ICARDA : International Centre for Agricultural Research in the Dry areas, Aleppo, Syria.

ICBP : International Council for Bird Preservation

ICBR : Incremental cost; Benefit ratio

ICEF : India Canada Environment Facility

ICF : International Crane Foundation

ICFOR : International Centre for Forestry Research

ICFRE : Indian Counil for Forest Research and Education

ICI : Indian Concrete Institute

ICIMOD : International Centre for Integrated Mountain Development, Kathmandu, Nepal.

ICID : International Commission on Irrigation and Drainage

ICIMOD : International Centre for Integrated Mountain Development, Kathmandu, Nepal.

ICIPE : International Centre of Insect Physiology and Ecology, Nairobi, Kenya.

ICITV : Institute de la carte International du Tapis vegetale

ICLARM : International Centre for Living Aquatic Resources Management, Metro Manila, Philippines.

ICMC	: International Congress on Metallic corrosion
ICPA	: International Congress in Plastics in Agriculture
ICRAF	: International Council for Research in Agro-forestry, Nairobi, Kenya.
ICRISAT	: International Crops Research Institute for Semi-Arid Tropics (Pattancheru, India)
ICSU	: International Council of Scientific Unions
IDB	: Inter American Development Bank
IDF	: Institute of Deciduous Forests (Jabalpur, India)
IDRC	: International Development Research Centre, Ottawa, Canada
IDS	: Institute of Development Studies
i.e.	: that is
IEES	: International Energy Evaluation System
IFA	: Indian Forest Act
IFAD	: International Fund for Agricultural Development, Cubae, Rome.
IFAP	: International Federation of Agricultural Produces, Paris, France.
IFC	: International Finance Corporation

IFDC : International Fertilizer Development Center, Muscale Sholas Alabama (USA)

IFGTB : Institute of Forest Genetics and Tree Breeding (Coimbatore, India)

IFOAM : International Federation of Organic Agriculture Movements

IFPRI : International Food Policy Research Institute, Washington, D.C., USA.

IFS : Indian Forest Service/International Foundation of Sciences, Sweden

IGF : Inspector General of Forests, New Delhi

IGFRI : Indian Grass land and Fodder Research Institute, Jhansi, U.P., India.

IGKVU : Indira Gandhi Krishi Vishwa Vidyalaya, Raipur 492 001, M.P., India.

IGNFA : Indira Gandhi National Forests Academy, Dehradun, India

IGNOU : Indira Gandhi National Open University, New Delhi

IGP : Indogangetic Plains

IHSA : Iodinated human serum albumin

IIAS : Indian Institute of Advanced Study, Shimla, India

IIASA : International Institute of Applied Systems Analysis, Luxemberg, Vienna, Austria

IIASE : International Institute for Aerospace Survey and Earth Sciences Enschede, Netherlands

IIE : International Institute of Entomology

IIED : International Institute for Environment and Development

IIFM : Indian Institute of Forest Management, Bhopal, India.

IIFS : Integrated Intensive Farming System

IIHR : Indian Institute of Horticultural Research, Bangalore, Karnataka, India.

IIM : Indian Institute of Management

IIMI : International Irrigation Management Institute (Sri Lanka)

IIP : Indian Institute of Petroleum

IIPR : Indian Institute of Pulses Research, Kalyanpur, Kanpur, U.P., India

IIRR : International Institute for Rural Reconstruction (Sillang, Philipines)

IIRS : Indian Institute of Remote Sensing, Dehradun, India.

IISR : Indian Institute of Spices Research, Calicut, Kerala; Indian Institute of Sugarcane Research, Lucknow, U.P., India.

IISS : Indian Institute of Soil Science, Bhopal, M.P., India.

IITA : International Institute for Tropical Agriculture, Ibdan, Nigeria.

IITM : Indian Institute of Tropical Meteorology

IJO : International Jute Organization

ILCA : International Livestock Centre for Africa/Addis Ababa Ethiopia

ILO : International Labour Organization/Switzerland

ILRAD : International Laboratory for Research and Animal Diseases Nairobi, Kenya.

ILRI : Indian Lac Research Institute, Namkum, Ranchi, Bihar, India.

IMD : India Meteorological Department

IMO : International Maritime Organization

IMTI : Irrigation Management Training Institute, Thuvakudy Trichi, Tamil Nadu.

In : Indium (element)

INIBAP : International Network for the improvement of Bananas and Plantains

INM : Integrated Nutrient Management

INRA : L'Institut Nationale de la Recherche Agronomique

INSAN : Institute for Sustainable Agriculture (Nepal)

INSFFER : International Network on Soil Fertility and Fertilizer evaluation on Rice, (IRRI)

IOBC : International Organization for Biological Control of Noxious weeds

IP : Indian Pharmacopia; Ion Product

I.P : Ion pairs

IPC : Indian Pharmaceutical Codex

IPCC : International Panel on Climate Change

IPI : International Potash Institute

IPIRI : Indian Plywood Industries Research Institute, Bangalore.

IPM : Integrated Pest Management

IPPC : International Plant Protection Convention

IPPS : The International Plant Propagators Society

Ir : Iridium (element)

IR : Infra Red

IRDC : International Research and Development Centre

IRDP : Integrated Rural Development Project

IRES : Integrated renewable energy systems

IRMDF : Institute of Rain and Moist Deciduous Forests, Jorhat, Assam, India

IRR : Internal Rate of Return
IRRC : Independent wire Rope Core
IRRI : International Rice Research Institute, Manila, Philippines.
IRTP : International Rice Testing Programme
ISAE : Indian Society of Agricultural Economics
ISCO : International Soil Conservation Organization
ISEE : International Society for Ecological Economics, U.S.A.
ISGIEC : Indian Sugar and General Industry Export Corporation Limited
ISGPB : Indian Society for Genetics and Plant Breeding
ISHS : International Society for Horticultural Science
ISI : Indian Standard Institution
ISNAR : International Service for National Agricultural Research, The Hague, Netherlands
ISO : International Organisation for Standardisation
ISRIC : International Soils Reference and Information Centre, Netherlands.
ISSS : International Society of Soil Science
ISTA : International Seed Testing Association, Oslo, Norway.
ISTF : International Society for Tropical Foresters

ISWT : Institute of Wood Science and Technology, Bangalore, India

IT : Internal Transition

ITCI : International Tree Crops Institute, Australia

ITTO : International Tropical Timber Organization, Japan

I.U. : International Units

IUCN : International Union for the Conservation of Nature and Natural Resources (Morges, Switzerland)

IUFRO : International Union of Forestry Research Organisation, Vienna, Austria.

I.V. : Importance value

IVI : Important Value Index (Vegetation analysis)

IVLP : Institute - Village Linkage Programme

IVRI : Indian Veterinary Research Institute Izatnagar, U.P., India.

IW : Irrigation Water

IWPD : Industrial Wood Production Division

IWRB : International Water fowl and Wetlands Research Bureau

IWRD : Industrial Wood Research Division

IWS : International Wool Secretariat Melbourne.

J

J : Joule; rate of heat storage in Canopy per unit ground area (w my^{-2})

JARI : Jute Agricultural Research Institute, Barrackpore, W. Bengal, India.

JMKV : Jawaharlal Nehru Krishi Vishwa Vidyalaya, Jabalpur 482 004, M.P., India.

JTRL : Jute Technological Research Laboratories, Calcutta, W. Bengal, India.

K

k : Boltzmann's Radiation constant (= 1.380×10^{-16} erg/deg)

k : Diseased trees (Class symbol V); reaction rate constant (Radio Isotope); Vonkarman's Constant (dimensionless); Relative crowding coefficient

k(a) : Dominant

K : Cumulus (cloud)

KAU : Kerala Agricultural University, Thrissur-680 654, Kerala, India.

k(b) : Dominant and suppressed

Kev : Kilo-electron volt

K : Turbulent transfer coefficient (m^{-2} s^{-2}); Potassium (element);

	Kelvin; Linear pair production attenuation coefficient (Radio Isotope)
kb	: Kilobases
KBAC	: Kernel Breadth After Cooking (Rice)
KBAN	: Kernal Breadth After Cooking (Rice)
KDC	: Kumaon Development Centre, New Delhi
KD_{50}	: Median Knock down dose. The dose required to cause knock down effect in 50 per cent of the test-insects
KFN	: Kerala Forest Research Institute Trichur, Kerala, India
Kg	: Kilogram
Kgf	: Kilogram force
Kj	: Kilojoule
KKV	: Konkan Krishi Vidyapeeth, Dapoli, 431712 Maharashtra.
KLAC	: Kernal length after cooking (Rice)
Km	: Kilometre
Km^2	: square kilometre
KN	: Cumulonimbus (cloud)
KNP	: Kaziranga National Park (Assam)/ Kanha National Park (Maharashtra), India.
Kp	: Pan coefficient
kPa	: kilopascal

Kr : Krypton (element)

KT_{50} : Median knock down time. The time required to cause knock down effect in 50 per cent of the test insects.

KVIB : Khadi and Village Industries Board, New Delhi.

KVIC : Khadi and Village Industires Commission, New Delhi.

KVK : Krishi Vigyan Kendra

kw : kilowatt

L

l : length; litre; levorotatory

L : Monin - obukhov length (m); Inductance (Radio Isotope)

La : Lanthanium (element)

LAD : Leaf Area duration

LAI : Leaf Area Index (m^2 m^{-2})

LAR : Leaf Area ratio

lat : latitude

lb : Pound (weight)

lb f : Pound force

L.C. grade thinning : Light Crown thinning

L.C. 50 : Median lethal concentration

LD50 : Median Lethal Dose

LDC : Least Development Country; Less Developed Country

LDI : Logging Development Institute, Dehradun, India

LEC : Land Equivalent Coefficient

LESS : Least Cost Estimation and Scheduling System

LER : Land Equivalent Ratio/Linear elongation ratio

LH : Luteinizing hormone

Li : Lithium (element)

LI : Lint Index

LLP : Lab-to-land programme

lm : lumen

LM : Lakkadive (Coconut Variety)

LMO : Light mineral oil

LO : Lakkadive Ordinary (Coconut Variety)

log : Base - 10 logarithm

Long : Longitude

LPD : Lowest Profitable diameter (logging)

LRAC : Long Run Average Cost

LRDC : Land Resources Development Centre, (U.K.)

LS : Latin Square Design

LSD : Least Significant Difference Test; Latin Square Design

LT_{50} : Median lethal time

Lu : Lutetium (element)

LVL : Laminated Veneer Lumber

LW : Low Volume

lx : lux

M

m (trees) : Dead and moribund trees (class symbol IV);

m : meter; mass of a substance, g or mg; Magnetic quantum number (Radio Isotope)

MPAU : Mahatma Phule Agricultural University, Rahuri, Maharastra, India.

mRNA : Messenger RNA

m^2 : square metre

m^3 : Cubic metre

M : Molar concentration; Isotopic mass of neutral atom, amu (Radio Isotope)

MAB : Man and the Biosphere programme (UNESCO)

mai : mean annual increment

MAI : Monetary Advantage Index

MAP : Manpower Allocation Procedure

MAR : Mean Annual Rainfall

MASER : Microwave Application by the Simulated Emission of Radiation

MAT : Mean Annual Temperature

MAU : Marathwada Agricultural University, Parbhani-431 402, Maharashtra, India

Max : Maximum
mb : Millibarn
m bar. : millibar
Mbf : 1000 board feet
MC : Moisture Content
MCA : Maximum Crown Area
mCi : Millicurie; Megacurie
MCI : Multiple Cropping Index
M.D. : Managing Director
MDE : Modulus of Elasticity
MDF : Medium Density Fibreboard
MDI : Moisture-Deficit index
ME : Metabolizable Energy
meq : milli equivalent
MeV : Million - electron volt
M.F.P. : Minor Forest Produce
Mg : Magnesium (element); Milligram
MIDS : Madras Institute of Development Studies, Chennai, India.
Min. : Minute (time); Minimum
MIO : Market Intervention Operation
smks : Meter - kilogram - Second
Ml : Milli litre
MLT : Multi Location Trials
mm : milli metre
MMA : Methyle Methacrylate
MMbf : 1 Million Board feet

Mn : Manganese (element)

Mo : Molybdenum (element)

MOE : Modulus of elasticity

mol. : mole

Mol. : Molecular

MOR : Modulus of Rupture

m.p. : melting point

MP : Marginal Product

MPAU : Mahatma Phule Agricultural University, Rahuri, Ahmednagar-413 722, Maharashtra, India.

mr : Milliroentgen

MR : Marginal Revenue

MRL : Maximum residue limit

MS : Mass Spectrometry; Murashige and skoog medium

MSSRF : M.S. Swaminathan Research Foundation, Madras-600 113, India.

MSP : Minimum Support Price

mt DNA : mitochondrial DNA

MU : Madras University, Chennai, Tamil Nadu, India.

MW : Molecular weight

N

n. : Haploid chromosome number; Neutron; observed number of random events (statistics)

2n : Zygotic (Somatic) Chromosome number

N : Newton; North; Normal (concentration); stocking (i.e.) number of trees per unit area; Nitrogen (element); Neutron number; Number of atoms; Normality; Numbers of individuals (Genetics)

Na : Sodium (element)

NAA : Napthyl Acetic Acid

NAARM : National Academy of Agricultural Research Management (Hyderabad) A.P., India.

NABARD : National Bank for Agriculture and Rural Development

NAD : Nicotinemide Adenine Dinucleotide

NADH : Nicotinemide Adenine Dinucleotide (reduced form)

NADP : Nictotinemide Adenine Dinucleotide Phosphate

NADPH : Nicotinamide Adenine Dinucleotide (reduced form)

NAFED : National Agricultural Cooperative and Marketing Federation of India

NAR : Net Assimilation Rate

NARP : National Agricultural Research Project

NARS : National Agricultural Research System

NAS : National Accounting System; National Academy of Sciences (Washington D.C., U.S.A.)

NATP : National Agricultural Technology Project

NAUA : Narendra Deva University of Agriculture and Technology, Faizabad 224 001, U.P., India

Nb : Niobium (columbium element)

NBAGR : National Bureau of Animal Genetic Resources, Karnal, Haryana.

NBFGR : National Bureau of Fish Genetic Resources, Shivanagar, Allahabad, U.P., India.

NBRI : National Botanical Research Institute, Lucknow, U.P., India.

NBSS & LUP : National Bureau of Soil Survey and Land-use Planning, Nagpur, Maharashtra, India.

N.C.A. : National Commission on Agriculture

NCC : Neutral Conservancy Council

Nd : Neodymium (element)

NDF : Neutral Detergent Fibre

NDP : Net Domestic Product

NDRI : National Dairy Research Institute Karnal, Haryana, India.

ND20 : Refractive index at 20°C

NDU : National Defence University (Washington D.C.)

NDUAT : Narendra Dev University of Agricultural Technology Faizabad, U.P., India

Ne : Neon (element)

NE : North East; Net energy

NERC : National Educational Resources Centre (New Delhi)

NF : Natural Forest

NFE : Nitrogen-Free Extract

NFL : National Fertilizers Ltd. (INDIA)

NFRC : Northern Forest Ranger's College, Haldwani (U.P.), India

NFTA : Nitrogen Fixing Tree Association, Hawaii, USA

NGRI : National Geographical Research Institute, Hyderabad, India

NGS : Normal Growing Stock

Ni : Nickel (element)

NIAEM : National Institute of Agricultural Extension Management Hyderabad, 500030, India

NIH : National Institute of Hydrology, Roorkee

NII : National Institute of Immunology (New Delhi)

NIN : National Institute of Nutrition, Hyderabad, Andhra Pradesh, India

NIO : National Institute of Oceanography

NIFTAL : Nitrogen Fixing Plants for Tropical Agricultural Lands (Hawai, USA)

NMNH : National Museum of Natural History, New Delhi

NMR : Nuclear Magnetic Resonance

NNRMS : National Natural Resources Management System

NOR : Nucleolar Organizing Region

NPL : National Physical Laboratory

NR : Nitrate reductase/Natural Rubber

NRC : National Research Council (Washington, USA)/National Research Centre on Orchids, Gangtok

NRCB : National Research Centre on Banana, Tiruchirapalli, Tamil Nadu, India

NRCC : National Research Centre on Camel, Jorbeer, Bikaner, Rajasthan, India

NRCPB : National Research Centre on Plant Bio Technology, New Delhi

NREP : National Rural Employment Programme

n.s.d. : no significant difference

NS : Nimbostratus (cloud)

NSDP : New State Domestic Product

nt : Newton

NVI : Nutritive value index

NW : North West

NWDA : National Water Development Agency, Ministry of Water Resources, New Delhi.

NWDP : National Wastê lands Development Board, Delhi.

NWMP : National Water Management Project

Ny : Normal yield

O

O : Oxygen (element)

OAU : Organization of African Unity

O.B. : Over bark

OCEES : Oxford Centre for the Environment Ethics & Society, U.K.

OECD : Organisation for Economic Co-operation and Development, Paris, France

OFI : Oxford Forestry Institute (U.K.)

OFT : On farm trial

OM : Organic Matter

Opt. : Optimum

opt. rot. : Optical rotation

ORP : Operational Research Project

ORSTOM : Office de la Researche Scientifique et Technique Outer Mer (France)

Os : Osmium (element)

OSB : Oriented Structural Board; Oriented Strand Board

OUAT : Orissa University of Agriculture and Technology, Bhubaneshwar 751003, India.

Oz : Ounce

P

P : momentum (mechanical unit); Proton

P : Phosphorous; Probability occurrence of a random event; Pressure (mechanical unit)

P : Parental generation; Probability; atmospheric pressure (pa); gross precipitation (m); Stumpage price; Weighted stumpage price reflecting qualitative differences in product; Probable error

P_1 : The parents of an F1

P_2 : The grand parents

Pa : Protoactinum (element); Pascal

p.a. : per annum

PAGE : Polyacrylamide gellecophoresis

p.a.i. : Periodic annual increment

PAL : Phenylalanine ammonialyase

PAN : Peroxy Acetyl Nitrate; Protective Area Network/Pesticides Action Network International Brussels, Belgium.

PAR : Photosynthetically Active Radiation; Potassium absorption ratio

PAU : Punjab Agricultural University, Ludhiana 141 004, India.

Pb : Lead (element); Net precipitation (m)

PB : Periodic Block

PBI : Protein Bound Iodine

PBR : Plant Breeder's Rights

PCARRD : Philippine Council for Agriculture Forestry, Fisheries and Natural Resources Research, (Los Banos)

PCCF : Principal Chief Conservator of Forests

PCP : Pentachlorophenol

Pd : Palladium (element)

PEG : Polyethylene Glycol

PEP : Phosphoenol Pyryvate; Progamme Evaluation Procedure

PER : Protein efficiency ratio

PF : Phenol Formaldehyde

Pft. : Phytochrome far-red form

PG : Philippains Green (Coconut Variety)

PGA : Phospholycenic acid

PGAI : Phospho Glyceraldehyde

PGS : Periyar Game Sanctuary, Kerala

PH : Hydrogen ion concentration

PKV : Punjarao Krishi Vidyapeeth, Krishnagar, Akola 444 104, Maharastra, India.

PLIM : Piling unutilized material

Pm : Promethium (element)
PMC : Pollen Mother Cell
Pn : Net precipitation (m)
PNW : Present Net Worth
Po : Polonium (element)
post-em : post-emergence
PPI : Potash and Phosphate Institute, Atlanta, Georgia, USA.
p.p.m : parts per million
ppmv : parts per million/volume
PQ : Photosynthetic quotient
Pr : Praseodymium (element)
PR : Participatory Research
PRB : Population Reference Bureau, (Washington)
PROSEA : Plant resources of South East Asia, Indonesia
PSI : Photosystem I
Pt : Platinum (element)
PUFA : Poly unsaturated fatty acids

Q

q : Quintal; specific humidity (kg kg^{-1})
Q : Quantity (electrical unit); production; biological yield; harvest; volume; Energy change in a nuclear reactor (Radio Isotope)
qt : quart

R

r : Regeneration category - Recruits Coefficient of correlation; cost of capital; diffusion resistance (s m^{-1}) Roentgen

R : Coefficient of Multiple correlation; Ranger; Economic rent; correlated counting rate; Nuclear radius (Radio Isotope); Radiation (W m^{-2}); Resistance (Electrical unit)

Ra : Radium (element)

RAP : Rapid Assessment Procedures

rad : Radiation Absorbed Dose

RAU : Rajasthan Agricultural University, Bikaner 334 001; Rajendra Agricultral University, Pusa, Samastipur 848 125, Bihar, India

Rb : Rubidium (element)

RBC : Red Blood Cell

RBD : Randomised Block Design

RBI : Reserve Bank of India

RCII : Relative Cropping Intensity Index

RCB : Randomized Complete Block Design

RCUES : Regional Centre for Urban and Environmental Studies

R.D. : Regional Director

Resp. : Respectively

Re : Rhenium (element)
Ref. Index : Refractive Index
rem : Roentgen equivalent man
rep : Roentgen equivalent physical
Rev. : Revolution (s)
RF : Resorchinol Formaldehyde
RGR : Relative Growth Rate
RGS : Rajaji Game Sanctuary, U.P., India.
Rh : Rhodium (element)
RH : Relative Humidity
Ri : Rechardson number (dimensionless); Root inducing
RIDF : Rural Infrastructure Development Fund
RLEGP : Rural Landless Employment Generation Programme
R.M. : Regional Manager
R.M. Value : Reichert-Malissel value of fat
Rn : Radon (element)
RNA : Ribose nucleic acid
r.p.m. : revolution per minute (Rotation per minute)
RQ : Respiratory Quotient
RRA : Rapid Rural Appraisal
RSNC : Royal Society for Natural Conservation
RSPB : Royal Society for Protection of Birds

RT : Relative turgidity

RTV : Rice Tungro Virus

Ru : Ruthanium (element)

RUI : Resource-use indices

S

s : second (time); spin quantum number; the standard deviation of a sample as computed from actual measurements; selection pressure

SAF : Saturated fatty acids

So : Original selfed plant

S_1 : First selfed generation (Progeny of S o Plant)

s trees : suppressed trees (Class symbol III)

S : Sub whippy shoots with a value of 0.25; siemens; stratus (cloud)

S+ : Sub whippy shoots with a value of 0.50

S : South; supply; indicator; site quality; Area; Sulphur (element); separation energy; stopping power; specific activity (Radio Isotope)

S_1,S_2,S_3 : The first, second and third generations respectively of selfed offspring; members of a series of alleles causing self-incompatibility

S_A : allowable cut derived supply function

SARC : Short-run Average Cost

SAARC : South Asian Association for Regional Cooperation

SAF : Society of American Foresters (Washington DC, USA)

SAPTA : South Asian Preferential Trade Arrangement

Sap-Val : Saponification Value

SAR : Sodium Adsorption Ratio

SARCD : Sweedish Agency for Research Cooperation with Developing countries

SAREC : Swedish Agency for Research Cooperation with Developing Countries

SAVE : SAARC Audio Visual Exchange

Sb : Antimony (element)

SBI : Sugarcane Breeding Institute, Coimbatore, India

Sc : Scandium (element)

SCA : Specific Combining Ability

SCAW : Scheduling and Controlling bag Automated Net Work System

SCI : Simultaneous Cropping Index

SCII : Specific Crop Intensity Index

SCO : Soil Conservation Officer

SCST : Society of Commercial Seed Technologists

SCU : Sulphur-coated urea

S.D. : Standard Deviation

S.D.I. : Standard Density Index

SDS : Sodium Dodecyl Sulphate

Se : Selenium (element); Soil expectation value

SE : South East; Standard Error

SEBI : Securities and Exchange Board of India

sec : Second

S/E : efficiency derived supply function

SEM : Scanning Electron Microscope; Standard Error of Mean

SEPASAT : Survey of Economic Plants for Arid and Semi-Arid Tropics (Kew, England)

S.F. : Social Forestry

SFRC : Southern Forest Rangers College, Coimbatore, India

SFS : State Forest Service

SFSC : State Forest Service College, Coimbatore, India.

S.G.1. : Seed directly produced by cross pollination and from which F_1 plants will grow

S.G.2 : Self bred seed from an F_1

S.G.3 : Self bred seed from an F_2

Si : Silicon (element)

SIDA : Swedish International Development Agency

SITE : Satellite Instructional Telecommunication Experiment

SK : Stratocumulus (cloud)

SKUAST : Sher-e-Kashmir University of Agricultural Sciences and Technology Dalgate, Srinagar 190 001 (in summer, May-October) and Gandhi Nagar, Jammu-Tawi (in winter, Nov. - April), India.

SLW : Specific Leaf Weight

Sm : Samarium (element)

SMIP : Special Minor Irrigation Programme

Sn : Tin (element)

SNF : Solids not fat

SNP : Shivpuri National Park, M.P., India.

Sol. : Solution(s)

Sol. Ash : Soluble Ash

Sp. : Species

Sp. Gr. : Specific gravity

SPAC : Soil Plant Atmosphere Continuum

SPCA : Society for Prevention of Cruelty to Animals

SPD : Split Plot Design

spp : Species Novea (New Species)

SPU : Serum protein uptake

SQ : Sturdiness Quotient

SR : Synthetic Rubber

Sr : Strontium (element)

SRD : Social Rate Discount

SRI : Stanford Research Institute (USA)

SRISTI : Society for Research and Initiative for Sustainable Technologies and Institutions

S.R.S. : Stratified Random Sampling

S.S : in a narrow sense

ssp : sub species

SSP : Soluble Sodium Percentage

S&T Input : Scientific and Technological input

Str. : Strain

Subf : Subforma (Subforms)

SW : South West

SWCS : Soil and Water Conservation Society, IOWA, USA

Syn. : Synonym

T

t. : time (s, min., hr.,); tonne; metric ton

tRNA : Transfer RNA

T. : Temperature (°C or ° K); age of the timber stand inclusive of regeneration lag; testsa; kinetic energy; Resolving time of a nuclear particle detector (Radio Isotope)

T-3 : Tri-iodothyronine

Ta : Tantalum (element)

TAC : Technical Advisory Committee of the CGIAR

TAFCORN : Tamil Nadu Forest Plantation Corporation Ltd., Tiruchirapalli, Tamil Nadu, India.

TANTEA : Tamil Nadu Tea Plantation Corporation Ltd., Coonoor, Tamil Nadu.

TARC : Tropical Agricultural Research Centre

Tb : Terbium (element)

Tc : Conversion period in years; Technetium (element)

TCA : Trichloroacetic Acid

TDN : Total Digestible Nutrients

Te : Tellurium (element)

TEM : Transmission Electron Microscope

Temp. : Temperature

TEPP : Tetraethyl pyrophosphate (Synthetic organophosphate insecticide)

TFAP : Tropical Forestry Action Plan

TFR : Total Fertility Rate

t&g. : tonged and grooved

TG : Thermogravimetry

TGR : Tree Genetic Resources

Ti : Titanium (element)

t/h : age of stand at harvest

Th. : Thorium (element)

Tl : Thallium (element)

TLA : Total Leaf Area

TLC : Thin-layer Chromatography

Tm : Harvest timing constraint; Thulium (element)

TMFTC : Tamil Nadu Forestry Training College, Vaigai Dam, Periyakulam Tamil Nadu, India

TMO : Technology Mission on Oilseeds, New Delhi

TMU : Tabacco Mosaic Virus

TNAU : Tamil Nadu Agricultural University, Coimbatore, India.

TNC : The Nature Conservancy

TNVASU : Tamil Nadu Veterinary and Animal Science University, Chennai, India

TOOL : Transfer of Technology for Development, Amsterdam, Netherlands

TPI : Tropical Products Institute, London, UK.

TRAFFIC : Trade Records Analysis of Flora and Fauna in Commerce

TREE : Training Research Education Extension

tr : Regeneration lag in years

TRI : Tropical Resources Institute, Yale University, U.S.A.

TRIPS : Trade Related Intellectual Property Rights

TSBF : Tropical Soil Biology and Fertility Programme

TSS : Total Soluble Solids

T&V System : Training and visit system

U

u : Regeneration category - whippy unestablished shoots with a value of 1.00; horizontal wind velocity along mean wind vector (X direction) (m s^{-1}); mutation rate

U : Uranium (element); Potential energy

u+ : Regeneration category - whippy unestablished shoots with a value of 2.00

UAS : University of Agricultural Sciences, Bangalore-560 065, Darwad-580 005, Karnataka.

U.b. : Underbark

UCIL : Uranium Corporation of India Limited

UF : Urea Formaldehyde

UHF : Ultra High Frequency

UHT : Ultra High Temperature

ULV : Ultra Low Volume

UN : United Nations (New York, USA)

UNCSTD : United Nations Conference on Science and Technology for Development

UNCTAD : United Nation Conference on trade and Development

UNDP : United Nations Development Programme

UNEP : United Nations Environment Programme (Nairobi, Kenya)

UNESCO : United Nations Educational, Scientific and Cultural Organisation (Paris, France)

UNFPA : United Nations Fund for Population Activities

UNIDO : United Nations Industrial Development Organization

UNRISD : United Nations Research Institute for Social Development

UNSO : United Nations Sudano-Sahelian Office

UNU : United Nations University (Tokyo, Japan)

UPOV : International Union for the Protection of New Varieties of Plant

USAID : United States Agency for International Development (Washington DC, USA)

USDA : United States Department of Agriculture

USDI : United States Department of Interior

USFS : United States Forest Service (Washington DC. USA)

USGS : United States Geological Survey

U.S.P. : United States Pharmacopea

UV : Ultra Violet

V

v : Velocity; veterans; win; mutation rate in a reverse direction

V : Present value; Vanadium (element); Volt; variance

Var. : Varieties (variety)

Vd : deposition velocity ($m\ s^{-1}$)

VFA : Volatile Fatty Acid

VHF : Very High Frequency

Viz. : Videlicet (Namely)

Vol. : Volume

VPKAS : Vivekananda Parvatiya Krishi Anusandhan Shala, Almora, U.P., India

Vs. : Versus

V/V : Volume/Volume

V/W : Volume/Weight

W

w : Vertical wind velocity (Z direction) (m s-1)

w : Regeneration category—Woody shoots

w+ : Regeneration category—Woody shoots browsed

W : West; wage rate; Wolfram or Tungsten (element) Watt; Work (Mechanical unit)

WARDA : West African Rice Development Association, Monrovia, Liberia.

Wb : Weber; Stored soil water

WB : World Bank (USA)

W.O.E. : Weed control efficiency

WCMC : World Conservation Monitoring Center

WEC : World Energy Conference; Wind Energy Conversion

WIHG : Wadia Institute of Himalayan Geology, Dehradun, India.

WII : Wildlife Institute of India, Dehradun, India

WK : Week

WLW : Wild Life Warden

WMO : World Meteorological Organization

WN : World Neighbours, Okalahoma, USA

w.p. : Wettable powder

WPA : World Pheasant Association

WPM : Woody Plant Medium

W.P. : Working Plan

WPO : Working Plan Officer

WPSI : Wild Life Protection Society of India, Delhi

WRI : World Resource Institute (USA)

wt. : Weight

WTO : World Trade Organisation

WUE : Water Use Efficiency

W/V : Weight/Volume

W/W : Weight/Weight

WWF : World Wild Life Fund
WWFN : World Wide Fund for Nature
WWI : World Watch Institute, Washington DC, USA

X

x : Absorber thickness, mg/cm^2; horizontal mean; wind direction; Basic number of chromosome in a polyploid series
X : access cost; Absorber thickness, cm (Radio Isotope)
X1 : First generation following irradiation with X-rays
Xe : Xenon (element)

Y

y : horizontal direction transverse to mean wind direction
Y : annual harvest volume calculated under the allowable cut rule
Yb : Ytterbium (element)
Yd : Yard
YSPUHF : Dr. Y.S. Parmar University of Horticulture and Forestry, Solan, Himachal Pradesh, India

Z

Z. : Vertical direction; Atomic number
Ze : Nuclear charge
Zn : Zinc (element)
Zc. : roughness length (m)
Zr. : Zirconium (element)
ZRS : Zonal Research Station
ZSI : Zoological Survey of India, Calcutta, India